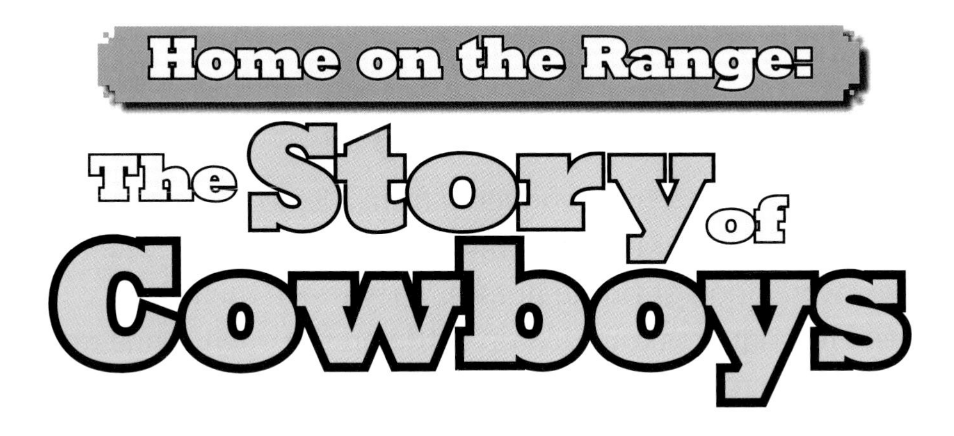

Printed in the United States of America

ISBN-13: 978-0-15-352851-4
ISBN-10: 0-15-352851-6

1 2 3 4 5 6 7 8 9 10 179 11 10 09 08 07 06

Visit *The Learning Site!* www.harcourtschool.com

The First Cowboys

The first cowboys were vaqueros from Mexico. *Vaquero* is a Spanish word that means "cowboy" or "buckaroo." The vaqueros were good at riding horses. They made their own saddles. The vaqueros rounded up cattle on the open range and herded them to ranches. They branded the cattle with the ranch brand. These symbols helped them tell their cattle apart from those of other ranches.

Early American settlers in the West learned from the vaqueros. They learned horse-riding skills. They learned how to rope cattle by using a lariat. They also learned how to use a branding iron.

At first, these people were called herders and drovers. Then they were called cowboys. That name stuck.

Tall Tales

Cowboys had lots of time to tell stories to one another. Some stories were about Pecos Bill. According to the stories, coyotes raised Pecos Bill. He rode a cougar and used a snake as a lariat. He wiped his mouth with cactus. He once even rode a tornado. They also say he taught cowboys how to brand cattle and throw a lariat. And he taught them to yell "Yippee!"

Cowboys often herded, or drove, cattle hundreds of miles.

People in the East were hungry for beef. There was plenty of it in the West, so ranchers raised more cattle, and they hired more cowboys.

It was up to the cowboy to get the cattle from the range to the train. In the fall, the cowboys rounded up the cattle from the open range. During the winter, the cowboys watched over the herds. In the spring, the cowboys herded, or drove, the cattle to towns that had railroads. These towns were called cow towns because of all the cattle there. The cattle were loaded onto trains and shipped east.

Some cattle drives could take many weeks because the cow town was hundreds of miles away. But by 1890, the railroad had reached many western towns. Cowboys no longer had to drive the cattle long distances.

Don't Fence Me In

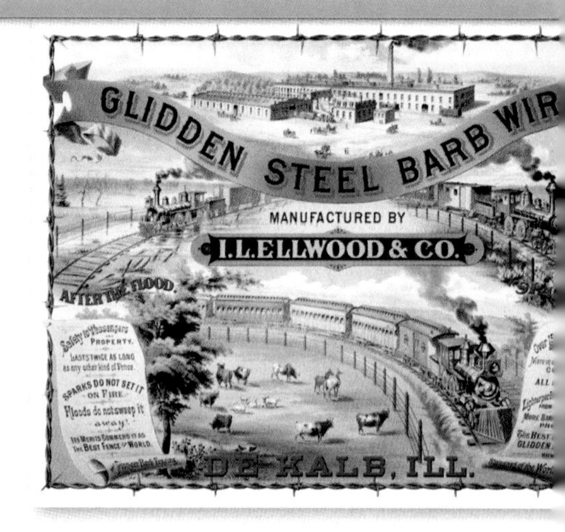

An advertisement for barbed wire

The West was open except for fences around a few ranches and farms. When more farmers and ranchers moved west, this open range became a problem. Farmers didn't want cattle running over their crops, so they built fences to protect their crops. Some ranchers built fences to keep people from stealing their horses and cattle. Most farmers and ranchers built fences out of barbed wire because stone and wood cost too much money.

Fences created conflict in the West. Some ranchers liked to use open lands for their animals, and they would cut the new fences. This started many conflicts between the farmers and the ranchers. By 1890, much of the open range was fenced.

Because of barbed wire, cattle could no longer roam the open range.

Fencing changed the jobs that cowboys did. They no longer had to round up cattle from the open range. Instead, they rode along the fences to check for holes and to repair any damage. They also rode around the ranch to make sure the cattle were safe and had enough to eat and drink.

Some ranches were so huge that cowboys rode far from the ranch house. At night, they lived in line camps. Some line camps were like small ranches with corrals and even a vegetable garden. Others were just tents pitched near the cattle.

Today, cowboys don't need to ride horses for days to check the fencing and pastures. They use trucks and four-wheel-drive vehicles, so line camps are not as common.

Herding cattle by helicopter

Cowboy Critters

Think about this question: Why weren't cowboys called "horseboys"? Without a horse, a cowboy couldn't get very far. Each day, the cowboy and his horse needed to cooperate to complete all of their demanding jobs.

In fact, a cowboy might need several horses for the job of rounding up cattle. Some horses were trained as cutting horses. That means they could cut, or separate, certain animals from the herd. This skill was important if a rancher's cattle ran with other cattle. The cowboy had to "cut" his boss's cattle from the herd.

Not all horses were tame, so broncobusting was a big part of cowboy life. A bronco is a horse that hasn't been trained.

A broncobuster rides a wild horse.

Cow-Dogs?

Dogs have also been used to herd cattle and other farm animals. One of the best "cow-dogs" is the Australian cattle dog. It has been bred for more than 150 years to work with cattle.

What kinds of cows did cowboys round up? The first cattle that American cowboys drove to market were longhorns. They were introduced by the Spanish hundreds of years ago. Their name refers to their horns, which could be from 3 to 5 feet across.

Longhorns could stay alive during cold winters and long drives to the railroad towns. But they were skittish, or fearful. They were skinny, and their meat didn't taste very good.

Longhorn cattle

When cattle ranchers fenced in their land, they tried new breeds of cattle. They no longer needed cattle strong enough to travel for many miles. Instead, they wanted cattle that were fatter, tasted better, and were not so skittish. Ranchers began raising shorthorn cattle.

Cowboy Things

Cowboys are known for the clothes they wore. The cowboy hat was based on the Spanish sombrero. Cowboys wore hats suited to where they lived.

In the northern ranges, the hat brim was narrower than brims in the Southwest. A hat with a wide brim would blow off in the northern wind! In the Southwest, cowboys needed a wide brim to keep out the sun and rain. A cowboy also used his hat as a drinking cup, a pillow, or a fan to cool himself and his horse.

Cowboys wore bandanas around their necks. A bandana is a folded piece of cloth. The cowboy used the bandana to keep out dust or water, wash himself, or cover his ears.

It was nice to relax and have someone else cook after a long day on the range.

Cowboys wore leather chaps over their pants. Chaps protected the cowboy from cactus thorns and sharp horns!

Cowboy boots were made for riding horses, not for walking. The boots were leather and had high heels. These heels helped the cowboy's feet stay in the stirrups of the saddle. Sometimes, cowboys wore spurs on their boots to make their horses run. They used a lariat to rope cattle and horses.

Cowboys carried a bedroll and other personal supplies. On most long cattle drives, a chuck wagon went along, and a cook made the meals. Horses pulled the chuck wagon. The wagon had a box on top for the kitchen.

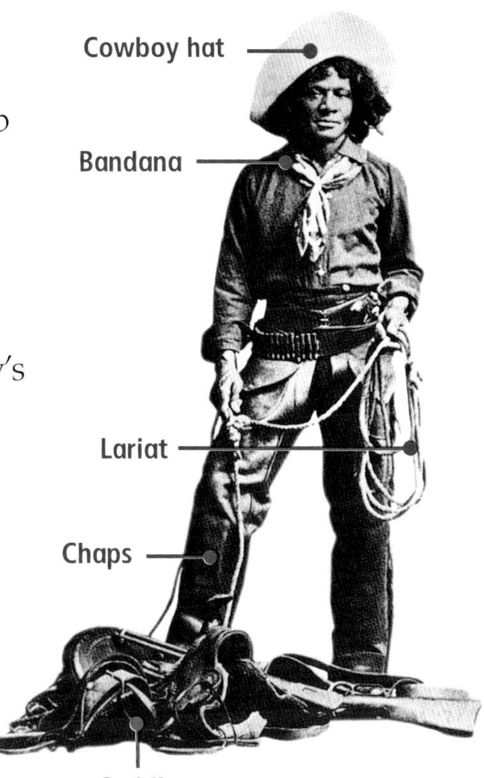

This photograph shows African American cowboy Nat Love. Cowboys' clothes were practical and gave them a special look.

Cowboy boots with spurs

A Hard Life with Danger

Cowboys who drove thousands of cattle hundreds of miles faced many hazards. Poisonous spiders and scorpions hid in boots and clothing. Rattlesnakes lived where cowboys traveled. One rattlesnake bite could kill a person.

Prairie dogs dug holes for their homes. A cowboy's horse could trip in a hole and throw the cowboy to the ground. Wild animals such as bears and mountain lions were a constant threat.

Cowboys had to herd cattle across deep, raging rivers. Horses and cowboys sometimes drowned. Cowboys had to fight prairie fires and survive raging blizzards and intense heat.

A great thing to say about a person of the Old West was "He'll do to ride the river with." When cowboys crossed raging rivers with cattle, they depended on each other for help.

Stampedes could end in disaster for cowboys and their cattle.

One of the biggest dangers for the cowboy was a stampede. Longhorn cattle frightened easily. The herd might suddenly start running if a cowboy sneezed. A stampede was dangerous. The stampeding animals could run over smaller and slower cattle. They could also trample the cowboys or their horses.

Cowboys had different ways to stop a cattle stampede. Sometimes, they would turn the cattle and get them to run in a tight circle. Sometimes, they gave up and let the cattle run. Then they'd have to round up all the cattle again.

To survive on the open range, cowboys had to cooperate. They worked together to take care of the cattle and to protect themselves from the dangers around them.

Break Time

Broncobusting, roping, and cutting cows were just a few cowboy skills. Sometimes, cowboys got together to show off their skills. They would compete to see who was the best rider, cutter, or roper. These events grew bigger and bigger because people liked to watch cowboys work. Soon, people began calling these events rodeos.

Buffalo Bill Cody started a Wild West show based on rodeos. It was filled with all kinds of cowboy events and performances. For his first show, he expected 100 cowboys. More than 1,000 cowboys showed up to compete for prizes. Soon, Buffalo Bill Cody's popular shows made the cowboy an American hero.

An advertisement for Buffalo Bill Cody's Wild West show

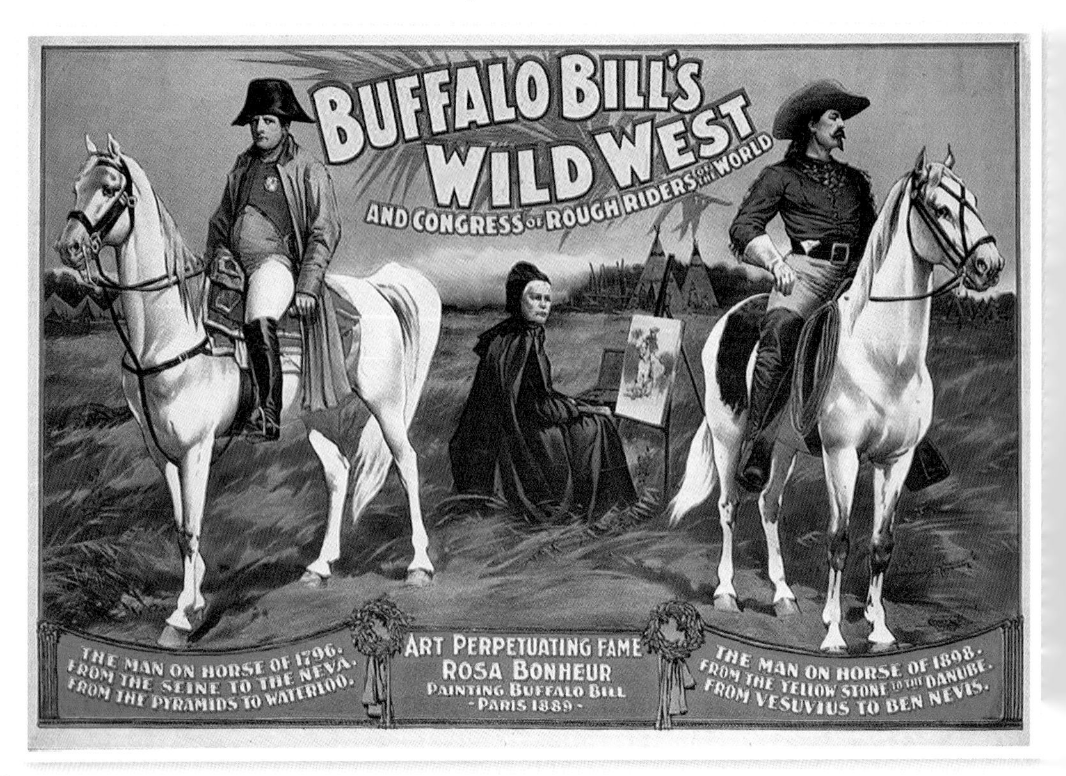

Cowgirls

The National Cowgirl Museum and Hall of Fame in Texas honors women who show the spirit of the American West. The term *cowgirl* wasn't used until rodeo days, but women were and still are a big part of the cattle business. Most of their jobs centered around the home, but women also worked alongside men on the roundups and cattle drives.

One rodeo event got its start when cowboy Bill Pickett leaped onto the horns of a steer to keep his horse from being hurt. He wrestled the steer to the ground and bit its upper lip. Pickett had once seen a bulldog do the same thing. This became a rodeo event called bulldogging. Today, this event is called steer wrestling, and biting is not allowed!

Rodeos are still important because cowboys and cowgirls need to practice their skills. Even though today's cowboys and cowgirls ride trucks or helicopters to take care of the ranch, they still use horses for some jobs.

Today, ranchers can use computers and machines to run some parts of a cattle ranch. But they still need skilled people to herd and care for the cattle.

Cowboys in Art and Music

Western art showed a way of life many people had never seen. Charles Russell was an artist who worked as a cowboy.

He knew how a cowboy lived and worked. His paintings of cowboys showed an exciting and sometimes dangerous life. Russell started selling his work and became one of the most famous western artists.

In the early days of the cowboy, people started writing stories about them. Cowboys became a big part of American folklore. One of the most popular writers was Zane Grey. Many of Zane Grey's books were made into movies. Cowboys became heroes on the big screen.

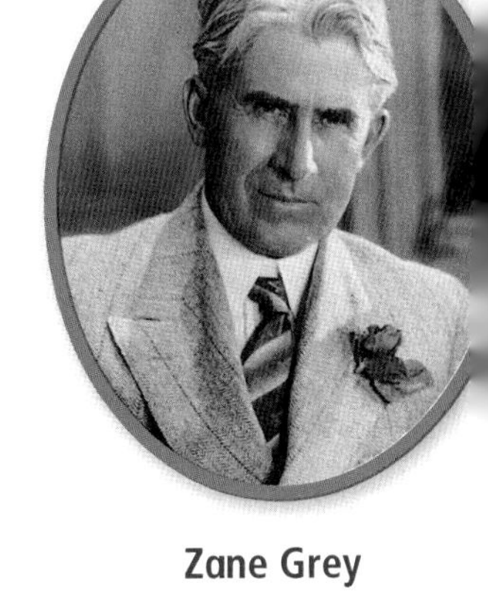

Zane Grey

This painting by Charles Russell shows cowboys rounding up cattle.

Movies made cowboy life look glamorous. Movie cowboys wore fancy, clean clothes. In real life, a cowboy's clothes were practical and usually dirty. In the movies, the cowboy's co-star was his horse. In real life, a cowboy needed several horses to get through the day. The rancher owned the horses. The cowboy owned only the saddle.

Cowboy songs were set to familiar tunes. Cowboys sang for each other around the campfire. Singing calmed the cattle, so they didn't stampede. Cowboys might also sing when they ran with a stampede. That way, a cowboy knew where his buddies were.

The song "Home on the Range" is one of the best-known songs from the Old West. In fact, country and western music can be traced back to the early days of the American cowboy.

Cowboys in the movies and on television often look very different from those in real life.

Think and Respond

1. Who were the vaqueros?
2. Why did farmers and ranchers fence the open range?
3. How did cowboys use their hats?
4. What were some jobs done by cowboys?
5. Why are cowboys seen as American heroes?

Activity

What was it like to be a cowboy or cowgirl in the Old West? With a small group, write a scene showing one of the jobs done by these adventurous people. Then role-play the scene for classmates.

Photo Credits
Front Cover Corbis; 3 William Albert Allard/National Geographic/Getty Images; 4 The Granger Collection; 4-5 Corbis; 5 Louie Psihoyos/Corbis; 6 Kevin R. Morris/Corbis; 7 (t) Paul A. Souders/Corbis, (b) Lowell Georgia/Corbis; 8 Corbis; 9 (t) Bettmann/Corbis, (b) David Stoecklein/Corbis; 10 John Carnemolla/Australian Picture Library/Corbis; 11 Mary Evans Picture Library/Alamy; 12 Prints and Photographs Division, Library of Congress; 13 The Granger Collection; 14 (t) Bettmann/Corbis, (b) Prints and Photographs Division, Library of Congress; 15 John Springer Collection/Corbis; Back Cover Prints and Photographs Division, Library of Congress.

Illustration Credits
2 Doug Knutson.